I walk to the pet shop with Dad. There, we talk to the clerk.

"Blake would like a pet," says Dad.

"What pet would you like?" the clerk asks.

That is a tough question! I *should* have an answer. *Usually* I pick things fast!

Should I get a snake? Would a skunk be a bit strange? This could be *rough!*

"What pets are here?" I ask.

"There are many!" the clerk answers. "We have fish, dogs, and frogs. We have skunks, skinks, and snakes."

This is rough! "Rough, rough!" I chant.

A "ruff, ruff" answers.

I talk to a pup. "Rough," I tell him.

"Ruff, ruff!" he says.

"Made your pick yet?" Dad asks.

"Yes, and it will not change."